NATIVE ORCHIDS OF SOUTH FLORIDA

BY

WALTER M. BUSWELL

CURATOR OF THE BUSWELL HERBARIUM

UNIVERSITY OF MIAMI

A BULLETIN of the University of Miami

Volume 19, No. 3, February 15, 1945

Coral Gables, Florida

NATIVE ORCHIDS OF SOUTH FLORIDA*

WALTER M. BUSWELL

In temperate regions all orchids are terrestrial while in the tropics many of them are epiphytic. Many of the terrestrial orchids found in South Florida are also found in northern Florida, some of them even as far north as Canada. Twenty five epiphytic orchids have been found in South Florida. *Epidendrum conopseum* found in the upper part of the state, is the only one not found in the more tropical region. On some of the epiphytic plants the flowers are small and inconspicuous, but there are few native plants that can show as wonderful a display of brightly colored flowers as the Oncidiums and Cyrtopodiums.

There are orchids in most of the hammocks in Dade County, upper Monroe County and in the Big Cypress hammock in Collier County where some of the rarest orchids are found. Parts of this hammock still in a natural state are the most fascinating spots in this entire region.

From the prairie we enter a green-roofed cathedral, containing column-like tree trunks decorated, often from base to top, with orchids and ferns, beautifully draped stumps and cypress knees, and an occasional woodland pool. During the summer months when there are frequent rains the slightly lower spots are more or less under water, but wading in this tropical forest is quite refreshing. At other times the ground is fairly dry in the low spots and since the standing water prevents the rank growth of vines and shrubs that are almost impassable on the higher ground, walking is much easier in the lower areas. Mammoth Royal Palms, the green plumes far above all the other trees add much to the attractions of the forest. Possibly only a botanist would be willing to crawl through thorny vines and dense shrubbery to gain admission to this cathedral, but to one with a love for the beautiful in nature it is well worth the trouble.

*South of Lake Okeechobee.

Publication sponsored by the Science Research Council of University of Miami.

On account of the annual fires and general destructiveness of man the most valuable natural attractions in South Florida are gradually disappearing and with many of the rare plants. Several orchids, both terrestrial and epiphytic, found here a few years ago, seem to have entirely vanished. At least, I have been unable to locate any of them for several years.

The Indians have discovered that some of the tourists will buy the larger orchids from them and these plants, once abundant, will soon be only a memory. It is particularly unfortunate for very few of them will be preserved by the tourist for more than a short time.

Orchids listed in Small's Manual of Southeastern Flora, not found in recent years are included in the following list:

Spiranthes polyantha	*Cranichis*
Basiphyllaea	*Triphora*
Centrogenium	*Brassia*
Prescottia	*Macradenia*

It is possible that many of these plants might still be found if I could locate them at the right time, as some of them resemble other more common species when not in bloom they could be overlooked. Apparently they were never very abundant and, being rare, were probably cleaned out whenever found. Some of them were destroyed by fires. In one of Dr. Small's early papers he mentions taking out several wagonloads of orchids and as he collected in the region for many years such collecting might account, in part at least, for the scarcity. Many commercial dealers in plants collecting truck loads of the most showy orchids have also helped in the extermination.

I am indebted to Professor Oakes Ames, Charles Schweinfurth and Dr. Donovan Correll of the Botanical Museum of Harvard University for corrections in the nomenclature and for many helpful suggestions.

Plate I

Top left: POGONIA OPHIGLOSSOIDES. *Right:* CALOPOGON PULCHELLUS.
Bottom left: EULOPHIA ALTA.
Right: CALOPOGON BARBATUS *var* MULTIFLORUS.

VEGETATIVE KEY

- Terrestrial plants
  - With leaves
    - Leaves basal
      - Leaves linear
        - CALOPOGON PULCHELLUS (1)
        - C. BARBATUS var. MULTIFLORUS (2)
        - C. PALLIDUS (3)
        - SPIRANTHES TORTILIS (20)
        - S. LACINIATA (21)
        - S. VERNALIS (22)
        - S. PRAECOX (23)
        - S. ODORATA (24)
        - HABENARIA NIVEA (27)
        - ONCIDIUM FLORIDANUM (42)
      - Leaves ovate, oval, elliptic, or lanceolate
        - CENTROGENIUM (9)
        - PRESCOTTIA (10)
        - CRANICHIS (11)
        - MALAXIS (12)
        - PONTHIEVA (13)
        - LIPARIS (14)
        - SPIRANTHES FLORIDANA (17)
        - S. GRACILIS (18)
        - S. LONGILABRIS (19)
        - S. CRANICHOIDES (25)
        - S. POLYANTHA (26)
      - Leaves like juvenile palm leaves
        - EULOPHIA ALTA (4)
        - E. ECRISTATA (5)
        - STENORRHYNCHUS (6)
        - BLETIA (7)
        - BASIPHYLLAEA (8)
    - Leafy stems
      - Leaves clasping, small, appressed — TRIPHORA (32)
      - Leaves not clasping
        - TROPIDIA (15)
        - HABENARIA REPENS (28)
        - H. QUINQUESETA (29)
        - H. STRICTISSIMA VAR. ODONTOPETALA (30)
        - ERYTHRODES (31)
        - ZEUXINE (34)
      - Vine
        - VANILLA PHAEANTHA (35)
    - One leaf near middle of stem — POGONIA (16)
  - Without leaves
    - Vine
      - VANILLA EGGERSII (36)
      - V. ARTICULATA (37)
    - Flowering stem only — CORALLORRHIZA (33)

- Epiphytic plants
  - Without pseudobulbs
    - Plants with leaves
      - Leaves two-ranked, sheathing at base — MAXILLARIA (49)
      - Creeping stems, stiff leaves
        - EPIDENDRUM RIGIDUM (51) — E. STROBILIFERUM (52)
      - Erect stems
        - Stiff leaves — E. ANCEPS (54)
        - Fleshy leaves — E. DIFFORME (50)
        - Leaves distant, not stiff — E. NOCTURNUM (53)
        - Several leaves on or near base
          - POLYSTACHYA (55) — IONOPSIS (56)
      - One leaf
        - PLEUROTHALIS (57) — LEPANTHOPSIS (58)
    - Plants without leaves, a mass of clinging roots
      - POLYRRHIZA (59) — HARRISELLA (61)
      - CAMPYLOCENTRUM (60)
  - With pseudobulbs
    - Pseudobulbs distinct
      - Pseudobulbs long, 6-12 cm. long, slightly flattened
        - BRASSIA (38) — EPIDENDRUM COCHLEATUM (47)
      - Small narrow pseudobulbs, 2-4 cm. long — MACRADENIA (39)
      - Pseudobulb large and hornlike — CYRTOPODIUM (40)
      - Pseudobulb round flattened, 2-4 cm. dia. — EPIDENDRUM BOOTHIANUM (45)
      - Pseudobulb pyriform, onion like — E. TAMPENSE (46)
      - Bulbs erect along the stems or roots — E. PYGMAEUM (48)
    - Pseudobulbs not distinct
      - Leaves broad, thick, stiff, 2-6 dm. long
        - ONICIDIUM LURIDIUM (41)
      - Leaves narrow
        - ONCIDIUM CARTHAGINENSE (43) — O. VARIEGATUM (44)

1. Calopogon pulchellus. (Plate I) (Limodoum tubersosum)
(Limodorum Simpsonii)

Plants 2-13 dm high; a broadly linear leaf from near base of flower stalk, 1-4 dm long, 3-30 mm wide.

Flowers in an open raceme of 5 to 22 large pink flowers, each 3 cm broad. Lip on upper side of the flower with thickened hair-like processes, fragrant. Capsule 2 cm long. February to June.

Usually in low wet grounds, some years abundant in certain fields, other years only a few plants. In marshes farther up the state they are found in much the same numbers every year. One of our most attractive wild flowers. The Florida plants are usually larger and with more flowers than those of northern states and in bloom earlier. Newfoundland to Florida.

2. Calopogon barbatus var multiflorus. (Plate I)
(Limodorum multiflorum)

Plants 2-3 dm tall with a very slender flower stalk and one or two linear leaves at base, 5 cm to 3 dm long, 1 mm wide, most of them not over 15 cm long, shorter leaves 2-8 mm wide, leaves often missing when the plants are in bloom, occasionally burned off, a few small bracts below the flowers. 3 to 15 rose-purple flowers in a terminal raceme, the flower about 2 cm broad. Lip bearded, middle lobe cuneate. The flowers open in quick succession and all may be open at the same time. Low pinelands on the West Coast. March. April. May.

Calopogon barbatus. (Limodorum parviflorum)

Similar to the other species, with pink or purple flowers, could possibly be found south of Lake Okeechobee but I have only found it farther up in the state.

3. Calopogon pallidus. (Limodorum pallidum)

Slender plants 3-5 dm tall. One linear leaf from near the base, 1-3 dm long, 1-3 mm wide. Often missing when the plant is in bloom.

Flowers in a terminal raceme of 6 to 10, or more, white and purple flowers, 2 cm broad, opening in slow succession, buds, flowers and fruits on the plant at the same time.

Pinelands, mostly on the west coast. March. April. May.

4. Eulophia alta. (Plate I) (Platypus altus)

5 to 9 dm tall. Leaves basal, broadly linear, 6-12 dm long. 6 cm wide, resembles leaves of Bletia or a seedling palm.

Flowers in a terminal raceme 3-6 dm long, the flowers large, green and brown or reddish-purple, greenish yellow in center, purple on edges. Lip with 3 short lobes, frilled or crested. Capsules 3-angled, 3-4 cm long, 12-18 mm wide. 3 wide flat ribs on the angles, 3 smaller ribs between. In or near hammocks. Spring to Fall.

South Florida. Mexico. West Indies. Central America and Northern South America. Africa.

5. Eulophia ecristata. (Triorchos ecristatus)

4-15 dm tall. Leaves broadly linear, 2-6 dm long, resemble seedling palm leaves. Flowers in a terminal raceme with long bracts. Sepals and 2 petals green. Lip brown, 3-lobed, 7-8 mm long. Capsule oval, 2 cm long. Pinelands. Summer and Fall.

6. Stenorrhynchus orchioides.

A tall stout plant to 11 dm tall. Basal leaves 1-3 dm long, 2-8 cm wide, often falling off before the flowers come on, more leaves may appear after the flowers. Upper leaves scale-like, sheathing. The whole plant colored much like the flowers.

Flowers in a dense terminal spike or raceme, red, green or yellow, 2 to 3 cm long. Lip lanceolate.

Pinelands and hammocks from Lee Co. north. Spring and Summer.

Mexico. Panama. West Indies. Peru. Brazil. Ecuador. Central America.

7. Bletia tuberosa. (Bletia purpurea)

Flowering stem 3-11 dm tall. One or several broadly linear leaves from one side of the bulb or corm. (flower stalk from other side)

Leaves 1-4 dm long. Resemble the leaves of a seedling palm and some other orchids. Flowers in a terminal raceme, the raceme sometimes branching. 10 to 20 rose-purple or reddish-purple flowers. Lip 10-12 mm long, 3-lobed, middle lobe frilled. Capsule 1-3 cm long, 5-8 mm wide, 3-ribbed. In rocky pineland March to June.

In hammocks on the west coast these plants were found growing in rotten stumps and logs, the round greenish bulb deep in the soft wood, leaves and flower stalk coming out of the stump some distance above the bulb but the leaves often dry or missing when the flower stalk comes out. Leaves 1-7 dm long, 3 cm wide. Flowers lighter colored than on the pineland plants and the spotted lip missing. Dr. Small did not believe me when I described the plants to him but he happened to be in Florida when they were in bloom and I took him to see them. Later he described them in Addisonia.

8. Basiphyllaea corallicola.

Plant 2-4 dm tall with clustered tubers. Basal leaves linear 2-7 cm long, sheathing scales above. Several rose-purple flowers in a terminal raceme, sepals 6-8 mm long. Lip broad, 3-lobed, middle lobe much larger. Found in shallow pockets in rocks in dry pinelands south and south east of Miami. Fall and Winter.

9. Centrogenium setaceum. (Pelexia setacea)

3-6 dm tall. Lower leaves oblong to elliptic, 4-10 cm long. Sheathing scales on upper part of stem. Flowers in a spike, greenish, lateral sepals linear-filiform 15-17 mm long. Lip 30-37 mm long with a long slender tip that is fringed on the sides. Flower with a long spur. Capsule 14-17 mm long. In leaf mold in hammocks. Dade Co. Fall and Winter.

10. Prescottia oligantha.

2-4 dm tall. Basal leaves broad oval, ovate or elliptic, 2-6 cm long. Sheathing scales above. Flowers small, in a terminal spike, white, pink or greenish. Petals 1 mm long. Lip 1-2 mm long. Capsule 4-5 mm long. In or near hammocks. Dade Co. Winter.

11. Cranichis muscosa.

1-3 dm tall. Leaves elliptic, oval or ovate, 2-7 cm long. Petioles long. Only sheathing scales on upper part of stem.

Flowers in a terminal spike, white. Petals 2-3 mm long, linear. Sepals ovate, 3-4 mm long. Lip oblong 3 mm. Capsule 1 cm long.

On stumps and cypress knees in cypress swamps and in moss around edge of lime sinks and potholes in hammocks.

Dade and Lee County. Winter and Spring.

12. Malaxis spicata.

Small plants with long fleshy roots, usually two basal leaves, oval or nearly orbicular, 2-10 cm long, 1-3 cm wide. Flowers in a terminal raceme on a slender stem, 1-3 dm tall. Flower small, white or greenish, lip orange in center. Florida to Virginia. As far south as Lee Co. on the West Coast but mostly farther north in the state. Low hammocks. Winter.

13. Ponthieva racemosa.

Plant 2-5 dm tall. Basal leaves large, oblong or lanceolate, sometimes flat on the ground or close to the ground.

An open raceme of greenish-white flowers on a nearly naked stalk. Flowers on penduncles about 1-2 cm long, standing out from the rachis. Whole flower nearly the same color, greenish-white with darker green lines. Hammock or open fields. Winter.

Florida. Alabama to Virginia. West Indies. Central America.

PLATE II

LIPARIS ELATA

14. Liparis elata. (Plate II) Twayblade.

1-4 dm tall. 4 to 6 large ovate or broadly elliptic leaves, thin, ribbed, pale green, 16-20 cm long, 8-10 cm broad, the broad petiole-like bases sheathing around the stem. Flowers in a raceme 12-15 cm long, each flower 1-2 cm wide, magenta or pale green splashed with magenta. Bracts at base of each flower green and magenta. Capsule oval, 1-2 cm long. Big Cypress in Collier Co. In rich humus on the ground or on rotten logs. July. August and other months. Mexico. West Indies. Central America. Brazil. Bolivia.

15. Tropidia polystachya.

3-5 dm tall. Stem branching, leafy. Leaves plicate, elliptic or lanceolate, thin, 5 to 7 parallel veins, narrowing to a long narrow sharp tip. Long sheathing bases. 5-20 cm long. 4-5 cm wide.

Flowers in a panicle, greenish-white. Petals 5-6 mm long. Lip 4-5 mm long, constricted above the middle. Capsule 9-14 mm long. Different months. "Open hammocks under shrubs, in Dade Co." Small.

16. Pogonia ophioglossoides. (Plate I)

Plant 1-5 dm tall. One lanceolate, oval or broadly elliptic leaf near middle of stem 1-8 cm long. 5-25 mm wide. A smaller leaf or bract just below the flowers. Large rose-pink flowers with fringed and crested lip, usually one to three flowers in Florida. In the northern states seldom more than one flower. Flowers raspberry scented. One of our most beautiful orchids.

Low wet places as far south as Lee Co. but more common farther north in the state. March to June. Newfoundland to Minnesota. South to lower Florida.

17. Spiranthes floridana.

Plants 1½-4 dm tall from a cluster of fleshy roots. Leaves short, elliptic, 1-5 cm long. 1½-2 cm wide. Smooth below the inflorescence. Spike slender, spiral, 3-12 cm long, flowers pale yellow and bracts of the raceme have a yellowish appearance.

Acid pinelands and marshes. Spring. Florida to South Carolina.

At least south to Broward Co. in Florida.

18. Spiranthes gracilis.

Plants 2-8 dm tall from a cluster of short thick roots. Leaves ovate, oval or elliptic. 1½-5 cm long. 1-2 cm wide. Spike spiral.

Flowers white, central portion of lip bright green. Spring.

I have found it in Lee Co. but Small lists it only to South Georgia.

19. Spiranthes longilabris.

Slender plants 3-4½ dm high with a cluster of fleshy roots.

Basal leaves lanceolate, 2-5 cm long. Spikes stout, spiral or secund, 6-10 cm long. Bracts very narrow. Flower white, 8-10 mm long.

Lip yellowish, toothed toward apex. Pinelands, Autumn.

20. Spiranthes tortilis.

Plants 2-6 dm tall from clusters of large roots. Glabrous. Basal leaves narrowly linear, thick, channelled on one side, 8-30 cm long, 2 mm wide. Often one or two leaves just above the base, scale-like on upper part of stem. Spike slender, 5-14 cm long, sometimes much twisted, others nearly straight and secund. Flower 4-6 mm long, white outside, pale green on inside, lip truncate, erose.

Rocky pineland. Dade Co. and Keys. Often abundant on Big Pine Key.

21. Spiranthes laciniata.

Plants 3-10 dm tall. Smooth below, densely pubescent above. Leaves linear, 1-3 dm long, 2-20 mm wide, mostly at base or a few on lower part of stem. Spike 8-24 cm long. Flower pubescent. Has been found near Miami and other parts of South Florida but apparently more common farther north in the state.

Oakes Ames says—"A bog or swamp orchid. In places where *Hypericum aspalathoides* and *Habenaria nivea* are found. Lip strongly fringed and the glandular hairs on flowers and rachis distinctly capitate. Last of May and early June when *S. praecox* and *S. vernalis* are mostly past blooming."

22. Spiranthes vernalis.

Plants 1½-6 dm tall from long fusiform roots. Long linear basal leaves extending for a short distance above the base. 5-10 mm wide. Upper part of stem pubescent. Spikes long, slender, 8-18 cm, often secund. "Rachis glandular pubescent. Lip thick, yellowish in center, distinctly glandular beneath. Hairs look dusty and are acute. Full bloom in April as *S. praecox* passes." Ames.

23. Spiranthes praecox.

Plants 2-8 dm high from a cluster of large white fleshy roots, the stem finely pale pubescent on upper part, glabrous below.
Leaves numerous, mostly at or near the base, linear. 1-2½ dm long, 4 mm wide. Small sheathing bracts on upper part of stem.
Flower spike 5-15 cm long, twisted. Flowers white, 6-8 mm long, frosty appearing on the outside. Lip smooth beneath, several fine green veins. Erose on edges. Not fragrant. In wet grassy pineland, marshes and in rock pits near pools of water. Starting to bloom in March.

24. Spiranthes cernua var odorata.

Plants 2-8 dm tall from slender fleshy roots. Basal leaves linear-lanceolate to 2 dm long, 1-2 cm wide, two or three leaves to middle of stem, large bracts above.

Large white flowers in spikes 5-10 cm long, 2 cm broad. Bracts conspicuously acuminate. Flowers very fragrant. Spring to Fall.

25. Spiranthes cranichoides. (Cyclopogon cranichoides)

Plants 1-3 dm tall. Basal leaves ovate or oval 2-6 cm long, satiny on upper side, pale green and slightly purplish beneath, sheathing scales on upper part of stem. Flowers in open terminal raceme, petals and sepals green, white and madder-purple. Lip 3-lobed, white 5-6 mm long. Hammocks. Winter and Spring.

26. Spiranthes polyantha. (Mesadenus lucayanus)

Plants 2-7 dm tall from a cluster of stout tuberous roots. Basal leaves ovate to elliptic, narrowing into petiole-like bases, 2-11 cm long. Flowers in a slender spike, greenish-purple, petals curved, 4-5 mm long. Capsule oval, ribbed, 4-6 mm long. Winter and Spring. Dade Co. and lower part of Elliott's Key.

27. Habenaria nivea. White Rein Orchid. (Gymnadeniopsis nivea)

2-6 dm tall. Two or three long linear leaves on lower part of stem, 5-30 cm long, 3-7 mm wide. Small narrow bracts on upper part of stem. Flowers in a dense terminal spike 6-14 cm long. The flowers small, white, with a spur 1 cm long. Sepals 4-6 mm long. Petals 3-5 mm. Lip linear, 5-7 mm long. May. June. July. Florida to Texas and New Jersey.

28. Habenaria repens.

Plants 2-5 dm high. Leafy. Leaves 3-15 cm long. 6 mm to 2 cm wide. Flowers in a dense terminal raceme, greenish, 1 cm wide. Spur slender, 9-13 mm long. An erect hood 5 mm broad with fine brown or dark red spots inside. Lip 5-7 mm long. Spring and Summer.

In ditches and small ponds, often floating in the water. Florida. Alabama to Virginia. West Indies. Central and South America.

Standley says "In Panama it has been found only on floating islands and logs in Gatun Lake and Gigante Bay".

PLATE III
Left: VANILLA EGGERSII. *Right:* HABENARIA QUINQUESETA.

29. Habenaria quinqueseta. (Plate III)

2-5 dm tall. Leafy. Leaves ovate or lanceolate, soft, smooth and glossy, clasping around the stem, middle leaves larger than others, 16-18 cm long, 2-4 cm wide. Basal leaves wider and shorter, 2-4 cm long. Small and narrow at top. Flowers in a terminal raceme 1-3 dm long. Flower about 3 cm broad, white or greenish white, the long narrow lobes widely spreading and resembling a large white spider. A slender green spur 3-6 cm long hanging down close to the rachis or twisting around it. Sometimes slightly fragrant.

June to Dec. Possibly other months. Pinelands. Florida to Texas and South Carolina.

30. Habenaria strictissima var odontopetala. (Habenella odontopetala)

A common terrestrial orchid in rich humus of hammocks, often on rotten logs, to 5 dm high. Leafy. Leaves lanceolate, thin, soft, smooth, light green, clasping the stem. Middle leaves larger than those above and below, the larger leaves 13 cm long, 3-4 cm wide, upper leaf 2-3 cm long. Flowers in a long terminal open spike or raceme of 2 to 25 flowers, the flowers green or greenish-yellow. Spur flat and often twisted, 20-22 mm long. Fragrant (like heliotrope?). Fall and Winter.

31. Erythrodes querceticola. (Physurus querceticola)

3 dm tall. Stem pale, smooth, with a conspicuous darker green ring at base of leaf. Leaves lanceolate, acute, base round, slightly fleshy, 4 cm long, 1-2 cm wide. Petiole 1 cm long, sheathing around the stem.

Flowers in a terminal raceme 5 cm long, or longer, the flower white, 4 mm long. Lip broad and notched on tip, a short stout pale fleshy spur 2-3 mm long. Bracts broad at base and partly clasping.

In humus. Big Cypress and hammocks south of Miami. Fall and Winter.

32. Triphora cubensis.

1-3 dm tall. From fleshy tubers. Leaves broadly ovate, 10-20 mm long, broad scale-like leaves clasping the stem and appressed.
Flowers nodding, from axils of leaf like bracts, magenta. Lip 3-lobed 8-9 mm long, middle lobe longer than others. Capsule drooping, 12-15 mm long. Pinelands, Dade Co. North part of Miami. July.

33. Corallorrhiza wisteriana. Coral-root.

2-4 dm high. No green color. Flowers in open racame 5-8 cm long. Lip 5-10 mm wide, white with reddish-purple spots, broad and round on tip, notched in center, a few fine sharp teeth on edge, or entire. Capsules 9-11 mm long. Winter and Spring. Hendry Co. and north.

34. Zeuxine strateumatica.

Plant 8 cm to 4 dm tall from a cluster of fleshy roots. Leaves dense, erect and close to stem, narrowly linear, partly sheathing at base, upper leaves often as long as the raceme, fleshy, revolute, only the prominent midrib noticeable beneath, green, or stem and leaves all reddish-purple, base of stem often red. 1-8 cm long. 3-5 mm wide. Flowers white, drooping, 5 mm long, one sepal and two petals together like an inverted spoon over the lip, one sepal on each side like a pair of wings. Lip yellow or greenish-yellow, 2-lobed at apex, 4 mm long. A long narrow bract at base of each flower. December to March.

In January 1938 I received a small plant that was collected on the Kissimmee prairie, something entirely new to me. Mr. Charles Schweinfurth of Harvard University identified it as this orchid, a native of China. It was first reported from Florida in 1936 and later from several widely separated localities in the state. No one seems to know how it was first introduced or how it is able to spread so rapidly but it may soon be found in every county.

It is less particular than any of our native orchids as it has been found in ditches, open fields and in dense grass on lawns, in sun or shade.

Plants were found on the perpendicular bank of a creek about 6 inches above the water under ferns and other plants, also in a lot that was in cultivation the year before, in dense weeds and grass.

35. Vanilla phtaeantha.

A vine with large round green stems running up into trees, branching and often forming a network around the tree trunk. They start from the ground, later separating near the ground and becoming epiphytic. Leaves lanceolate or broadly linear, 10-18 cm long, 2-3 cm wide. Flower tubular, white, 7-8 cm long, 15-18 mm wide. Sepals linear, green, 6-8 cm long. June-July. Big Cypress.

36. Vanilla eggersii. (Plate III)

Vine with long stout green stems 1-2 cm diameter. Leaves small and bract-like, distant and not conspicuous.

Flower cylindrical, white, purple and yellow. 2-3 cm long. Lip curved down, purple or violet, frilled or crisped, a row of short yellow raised spurs through center of lip, tube creamy yellow below, greenish white above, reddish-purple at mouth and with purple spots down in the tube. Very fragrant. Capsule slender, clavate, 5-7 cm long. Spring. Summer. Abundant on high dry rocky ground on Big Pine Key running along the ground and over shrubs. Also on some other Keys.

37. Vanilla articulata.

Vine with stout stems similar to *V. eggersii,* no leaves. Said to be found in land subject to overflow part of the year. Flowers similar to *V. eggersii.* I have not found this species. Mainland swamps and low lands. Cape Sable region.

38. Brassia caudata.

Large flattened pseudobulbs, 6-14 cm long. Two broad elliptic leaves from apex of bulb, 2-4 dm long, 2-4 cm wide. Bulbs and leaves resemble those of *Epidendrum cochleatum* but on *E. cochleatum* the flower stalks are from tip of bulb, from base on *Brassia.* Flowers in a loose raceme, light yellow with brown spots, sepals long, narrow, 6-14 cm long. Capsule 3-angled, 3-ribbed, beaked on both ends, 5-8 cm long. January to July. Florida. Mexico. West Indies and farther south.

39. Macradenia lutescens.

Small narrow pseudobulbs 2-4 cm long, about same width as the petiole, the single narrow elliptic leaf from tip of bulb, leaf similar to that of *Brassia,* but smaller. 5-6 cm long, 1-3 cm wide. Flower stalk from base of bulb, the small yellow inconspicuous flowers in an open raceme. Capsule 2-3 cm long. Blooming at different times in the year.

Small says "Low down on trunks and limbs of trees in hammocks".

40. Cyrtopodium punctatum. (Plate IV)

Long hornlike pseudobulbs with narrow dark rings 2-4 cm distant. The bulbs 2-6 dm long, 1-4 cm thick, in large clusters on old plants, bulbs, leaves and flower stalks all from a dense mass of brownish roots,

Plate IV

Top: ONCIDIUM LURIDUM. *Bottom:* CYRTOPODIUM PUNCTATUM.

plants forming clusters sometimes over three feet wide. Leaves linear-lanceolate, from tip of bulbs, 3-7 dm long, 1-5 cm wide, 2-ranked, drooping toward tips.

New leaves appear first, then the flower stalks from base of bulbs. The flowers in a large widely branching panicle, abundant and showy. Flowers like a mass of frilled and twisted ribbons thickly mottled yellow and reddish-purple or madder-brown, each 5 cm broad.

Sepals greenish yellow marked with oblong madder-brown spots, margins strongly undulate. A long wavy-edged pale green lanceloate bract at base of each pedicel with reddish-purple spots like the spots on the flowers. Lip 3-lobed, yellow, purple and brown. March to May.

On trees, logs or stumps in hammocks. Once abundant but as the Indians are permitted to sell the plants along the highway they will soon be as rare as several other once common orchids.

41. Oncidium luridum. (Plate IV) (Oncidium undulatum)

Pseudobulbs 1-2 cm long, not conspicuous. Leaves large, thick and stiff, oblong-lanceolate or broadly elliptic, to 6 dm long.

Flowers in a widely spreading panicle 1-2 m long. The flowers variable on different plants. Pale green or olive-yellow spotted with brown, the center brown or maroon with very little yellow. Or olive-yellow with madder-brown spots, center yellow. About 5 more combinations of color. Plants with darker colors seem to be more common. Edges of petals crisped or frilled. Old plants with three or four flower stalks may have over 1000 flowers at one time.

Capsule 5-8 cm long, 2 cm broad. 3 broad flat ribs with 2 rounded ribs between each. Flowers March to June. Mostly in Cape Sable region.

42. Oncidium floridanum.

Pseudobulbs on old plants resemble small bulbs of Cyrtopodium but on many plants they are not evident. 8-12 cm long. 3 cm thick.

Leaves linear, 3-10 dm long, 2 cm wide, somewhat 2-ranked at base, often from the ground with no bulbs in sight. Flower stalk slender, 1-2 m long with several branches on upper part 2-8 cm long, the flowers scattered along these branches, usually only one to three open at one time on each branch. Flower small, yellow with small brown spots, 2-4 cm broad. April to July, usually after *O. luridum* is past blooming.

Capsule 2-3 cm long, 8-10 mm wide, 3-angled with a small rib between each of the angles. Said to be epiphytic but I only find the plants growing in rich humus on the ground or on a rotten log.

South Florida. Mexico. West Indies. Central and South America.

43. Oncidium carthaginense.

Pseudobulb very small. Leaf narrowly elliptic, 9-30 cm long. Flower stalk 5-15 dm long. Flowers pink, mottled with magenta. Hammocks, Cape Sable region. Not found recently

44. Oncidium variegatum.

Pseudobulbs minute. Leaf curved, purplish on under side, margins scabrate, 3-7 cm long, 3 mm broad. Flower stalk 10-35 cm long, the flower white and green, purple and brown. Sepals pale green or white, barred with purple. Petals orbicular, white, magenta and purple. Lip 3-lobed, white and yellow, 1 cm long. Hammocks of South Florida and West Indies.

45. Epidendrum Boothianum.

Pseudobulbs round, flattened, 2-4 cm diameter, often in dense clusters standing on edge like a lot of green coins. 2-3 leaves from tip of bulb, narrow elliptic, thin, not stiff, 6-12 cm long, 2-3 cm wide. Flowers in open raceme, 2-3 cm broad. Sepals and two petals pale green with brown spots, sepals 12-15 mm long, 5 mm wide, two petals very narrow, otherwise like sepals. Lip green or yellow, 3-lobed. Flowers slightly fragrant. August to February. Cape Sable region and upper Keys, on living or dead trees and stumps.

46. Epidendrum tampense.

Pseudobulbs pyriform, like a small onion, usually in clusters, green or purplish. Leaves linear, thick, rigid, 9-30 cm long, 8-20 mm wide. A single leaf, or several leaves, from apex of each bulb, the flowering stalk also from the apex, long and slender, 1-9 dm long. Flowers in a widely spreading panicle, each flower 3-4 cm broad, sepals greenish yellow splashed with reddish-brown or magenta. Lip white or yellow, stained, striped or suffused with magenta or with a madder-purple spot in center, the flowers variable in color and the most attractive of the smaller plants, also one of the most common. Capsule 3-lobed, 2-4 cm long. April and May, occasionally later.

PLATE V

Top: LEPANTHOPSIS MELANANTHA. *Bottom:* MAXILLARIA CRASSIFOLIA

47. Epidendrum cochleatum.

Long bright green pseudobulbs, 6-12 cm long, 2-3 cm wide. Long linear leaves from apex of bulb, 1-4 dm long, 1-3 cm wide. Flower stalk from tip of bulb between the leaves, open raceme of several flowers. Sepals and two petals long and narrow, in a circle around the lip, pale yellow. Lip spoon-shaped, 15-18 mm diameter, purple or partly white with darker purple stripes or veins, tubular base of flower 1 cm long, green with purple spots on one side, pale greenish-yellow on other side, purple around the base. Capsule winged. 2-3 cm long. Blooming at different times in the year. Common in most hammocks.

48. Epidendrum pygmaeum.

Stems creeping and branching, to 3 dm long. Many long white roots. Small fusiform pseudobulbs 2-4 cm long, erect and 2-4 cm apart on the stem. Leaves elliptic-oblong, acute, 1-4 cm long, 10-13 mm wide. Flowers at tip of pseudobulbs. Sepals and petals greenish. Lip 4-5 mm wide, white with a purplish spot. Capsule 12-15 mm long. Winged. Hammocks. Spring.

49. Maxillaria crassifolia. (Plate V) (M. sessilis)

Described as having pseudobulbs but they are not noticeable. Leaves 2-ranked, sheathing at base, lanceolate, thick and stiff, 2-3 dm long, 1-2 cm wide. Flower from near base of the sheathing leaves, barely protruding from under the edge of leaves, cylindric, pale yellow, 6 lobes straight out, or nearly so, the small mouth open just enough to show the stigma. 1-2 cm long. Capsule oblong, 2 cm long, 8-10 mm wide, pale green, wide flat ribs. Not uncommon in the Big Cypress on trees and logs, usually several plants occuring where any of them are found. Florida. Jamaica. Cuba. Mexico. Venezuela. Brazil. Costa Rica.

50. Epidendrum difforme.

Stout light green leafy plants 1-3 dm tall. Leaves oblong, thick and fleshy, larger leaves in the middle, 3-8 cm long, 1-2 cm wide, round or slightly notched on tip, sheathing around the stem to next leaf below,. Flowers in umbel-like clusters, several greenish-white or pale green flowers 2-3 cm broad, 5 narrow spatulate lobes spreading like spokes of a wheel. Sepals 15 mm long, 4 mm wide, 2 petals narrower. Lip standing out in front on a column 8-10 mm long, broad reniform, glistening white, 7-8 mm long, 15 mm broad, slightly fragrant.

Capsules oblong, bright green, soft and velvety in appearance, 15-25 mm long 6-ribbed. Flowers usually in April. May. Aug. to November. Hammocks.

51. Epidendrum rigidum.

Creeping plants, often with a long row of erect stems from the creeping branches. Long, grayish, often dense roots. Leaves oblong or linear-elliptic, thick, rigid, on two opposite sides of the plant, sheathed at base to next leaf below, larger leaves at top, tip round, obtuse or notched, 2-8 cm long, 7-15 mm wide.

Flowers in slender terminal racemes from axils of large bracts, small, greenish and not conspicuous. Rachis flattened, edges sharp.

Capsule 13-18 mm long, oval, sessile. Winter and Spring.

52. Epidendrum strobiliferum.

Small slender creepng plants with one or several erect branching stems. Slender grayish roots. Old leaf bases sheathed around the stem. Leaves linear, stiff, obtuse or notched on tip, midrib and edges often dark purple, sheathing at base to next leaf below.

Several small white flowers in a terminal raceme from axils of small bracts, the flower 4-5 mm broad. Sepals lanceolate 4 mm long. Lip ovate-cordate. Capsules oval, about 1 cm long, faintly ribbed, a brown bract on one side enclosing about one third of the capsule. Flowers in fall, occasionally in winter and spring.

"This species first discovered in the U. S. in 1904 near Naples on the west coast where it was found growing on a large Sweet Bay Tree. Later another was found near Everglades." Ames.

South Florida hammocks. West Indies. Central and South America.

53. Epidendrum nocturnum.

Tall straight stems 3-11 dm high from a cluster of long roots.

Leaves distant, broadly linear, round and slightly notched at apex, stiff and somewhat fleshy, clasping around the stem to next leaf below. 8-16 cm long, 2-3 cm wide. Rachis more or less zig-zag with a pale clasping bract at each angle. Flowers from axil of each bract, narrow yellowish sepals and two petals 4-6 cm long, 2-3 mm wide in a circle like spokes of a wheel. The lip straight out from center. A cylindric column, white on outer two thirds, greenish-white at base, middle lobe 3-4 cm long, bright yellow, two lateral lobes broad and round at base, narrowing to tip, much like a pair of bird wings. Lip pure white except for a small yellow spot. An odd and very attractive flower. Fragrant. Capsule oval or elliptic, 4-5 cm long.

Dr. Small says "Flowers mostly green". The lip is mostly white, sepals and petals yellowish. Spring and several months in the year.

54. Epidendrum anceps.

Plants 2-7 dm tall, the stems often tufted. Leaves close, alternate on two sides, oblong-lanceloate, 5-25 cm long, stiff, dull green, often partly or entirely reddish-purple, leaves sheathed around the stem to next leaf below. Upper leaves larger, gradually smaller to base.

Flowers in umbel on a long slender flattened sharp-edged terminal peduncle. Flower bronzy yellow, flat, about 2 cm wide. Sepals oblong or spatulate, 2 petals narrower. Lip spoon-shape, standing out from other parts of flower, a greenish tube with a broad 3-lobed tip, a purple spot at end of tube, or column. Capsule oblong, ribbed, 15-20 mm long. Blooming several months in the year. One of the common Epidendrums. Hammocks of South Florida. West Indies. Mexico. Central and northern South America.

55. Polystachya luteola. (Polystachya minuta)

A dense mass of pale green roots spreading over the tree trunk or branch clinging to the bark, often with many flowering stalks 1-6 dm tall. Usuallly 2-3 oblong leaves clasping the stem, 5-20 cm long, 2-4 cm wide, thin, abruptly pointed.

Flowers in a terminal panicle, white, turning yellowish, 5-10 mm wide. Sometimes the flowers are all on one side of the flattened, sharp edged rachis. The flowers from axils of broad and partly clasping green bracts resemble small white or greenish-yellow butterflies with partly folded wings. March to November. Capsule 9-12 mm long, 5 mm wide, ribbed. Often in dry oak hammocks with *Epidendrum tampense.* The only two commonly found in such places.

56. Ionopsis utricularioides.

Plants 1-6 dm tall. Fine gray roots wrapped around the small branches and twigs of trees, mostly toward the ends of branches. Leaves thick and stiff, finely ribbed on the upper surface, often purplish, linear, 1-2 dm long, 102 cm wide. Flowers in a widely branching panicle on long purplish peduncles, the panicles 2-3 dm long, 1 to 2 dm wide. Lavender

PLATE VI

Top: HARRISELLA PORRECTA. *Bottom:* POLYRRHIZA LINDENII.

or purple, occasionally white, lip 1 cm long, 12 mm wide, deeply notched on tip, sides crenulate. Other parts of flower lavender with fine purple stripes. Pistil yellow and purple. Capsule triangular, 1-3 cm long, a long slender spur on tip.

57. Pleurothallis gelida.

1-3 dm tall. One broad thick oblong or elleptic leaf 8-16 cm long. Flowers in a slender raceme from the axil of a bract, appears to be from side of petiole, somewhat bell-shape with 3 spreading lobes, 8-10 mm long, pale yellow, pubescent on inside, fragrant (hyacinth odor?).

Capsule 8-10 mm long, greenish-yellow, ribbed, when dry it turns inside out and is then covered with long pure white hairs. Winter. Spring. Big Cypress.

58. Lepanthopsis melanantha. (Plate V) (Lepanthes Harrisii)

Small plants with spreading roots and a single oval or elliptic leaf 1-2 cm long. Flowers in short racemes, very small, a glass needed to study them, reddish-purple, 1-2 mm long. Capsule oval, 6-ribbed, 3-4 mm diameter. Spring. Summer. Big Cypress.

Fawcett and Rendle found it in Jamaica at 2000 ft. altitude.

59. Polyrrhiza Lindenii. (Plate VI)

Plant a network of long slender green roots spreading for several feet over trunks and branches of trees. Flower pure white, a broad lip 3-4 cm long, 1-2 cm wide with a pair of twisted lobes at tip like a pair of Ram's horns, lobes 5-7 cm long.

A white spur 12-15 cm long, the slender peduncle 8-14 cm long. Flowers fragrant. Capsule 5-8 cm long. The odd pure white flowers make this one of the most attractive of the epiphytic orchids. April to July. Small says "on palms" I find them on many different trees but not on palms.

60. Campylocentrum pachyrrhizum. (Plate VII)

Long flat shining dark green roots clinging tightly to the bark and winding around the trunks and branches. Roots wider and darker green than on *Polyrrhiza*. 5-8 mm wide. Flowers in a spike, greenish-yellow, small and inconspicuous, the green roots slightly raised and free from the host plant under the flowers and fruit. Capsules clustered in a double row on a slender bracted peduncle, oblong, olive green with brown papery tips. 5-6 ribs. 7-8 mm long. Aug.-Sept. to Spring. Big Cypress. Collier Co.

PLATE VII
CAMPYLOCENTRUM PACHYRRHIZUM

61. Harrisella porrecta. (Plate VI)

The slender gray roots wrapped around small branches of trees. Flowers in small racemes on very slender flower stalks, the tiny greenish-yellow flowers 2-3 mm long, inconspicuous.

Capsule oval or round, about 5 mm diameter, more conspicuous than the flowers.

Small says "Often on trunks of conifers, *Sabina* and *Taxodium*." Ames "On Junipers, *Acer rubrum, Cephalanthus* and *Fraxinus caroliniana*".

I find it on Citrus trees with *Ionopsis*.

Burmannia Family. Burmanniaceae.

Small plants related to Orchids. Flowers regular. (Orchidaceae, Flowers irregular.)

62. Burmannia biflora.

Very slender bluish white stems 5-15 cm tall. A few minute scales mostly on lower part of stem. One to three small terminal flowers, the winged corolla white or pale blue. Capsule 4-5 mm long.

In low grassy pinelands, only observed when standing directly over the plants and if they are in bloom. Spring to Fall and early winter.

63. Burmannia capitata.

Slender stems 5-20 cm tall, simple or occasionally with two stems from near the base or on upper part of stem. Small distant scale-like leaves 1-5 mm long. Flowers in dense head-like clusters of several small cylindrical white or purplish flowers 3-4 mm long. Capsule 2-3 mm long. Low wet prairie. Lee & Dade Co. and north.

64. Apteria aphylla.

Stem simple or branching, very slender, 5-20 cm high. Scales 1-2 mm. Two to six distant nodding flowers on upper part of stem, 10-15 mm long, on pedicels 3-10 mm long. Flower cylindrical, white or purple, lobed at apex. Capsule oval or ovate, 5-8 mm long.

Stem extending into the ground as a long straight or twisted tap root similar to the stem above ground, but paler. Sometimes longer than the stem. Hammocks. Spring to Fall.

Printed in the USA
CPSIA information can be obtained
at www.ICGtesting.com
LVHW011056260124
770059LV00003B/41

9 781013 416859